Elapsed Time

A complete workbook with lessons and problems

By Maria Miller

ISBN 978-1523233663

EDITION 7/2016

Contents

Preface

Hello! I am Maria Miller, the author of this math book. I love math, and I also love teaching. I hope that I can help you to love math also!

I was born in Finland, where I also grew up and received all of my education, including a Master's degree in mathematics. After I left Finland, I started tutoring some home-schooled children in mathematics. That was what sparked me to start writing math books in 2002, and I have kept on going ever since.

In my spare time, I enjoy swimming, bicycling, playing the piano, reading, and helping out with Inspire4.com website. You can learn more about me and about my other books at the website MathMammoth.com.

This book, along with all of my books, focuses on the conceptual side of math... also called the "why" of math. It is a part of a series of workbooks that covers all math concepts and topics for grades 1-7. Each book contains both instruction and exercises, so is actually better termed *worktext* (a textbook and workbook combined).

My lower level books (approximately grades 1-5) explain a lot of mental math strategies, which help build number sense — proven in studies to predict a student's further success in algebra.

All of the books employ visual models and exercises based on visual models, which, again, help you comprehend the "why" of math. The "how" of math, or procedures and algorithms, are not forgotten either. In these books, you will find plenty of varying exercises which will help you look at the ideas of math from several different angles.

I hope you will enjoy learning math with me!

Introduction

Elapsed Time Workbook contains lessons and exercises suitable for grades 3 and 4.

The first lesson, "How Many Hours Pass", uses a number-line diagram to help students figure out how many complete hours pass from one given time to the other, including if there is a change from AM to PM.

The lesson "How Many Minutes Pass?" encourages students to use a practice (analog) clock or visualize the minute hand moving along the clock face to figure out the elapsed time. It also explains the subtraction method for easy time intervals.

The next two lessons about calculating elapsed time emphasize dividing a time interval into parts that are easily calculated. For example, to find the elapsed time from 10:30 AM to 7:00 PM, the student learns to find the elapsed time from 10:30 AM to 12:00 noon, and then from 12:00 noon to 7 PM. The same principle is followed when the time-interval seems more complex.

Lastly, the lesson "Elapsed Time or How Much Time Passes", introduces the concept of adding or subtracting hours and minutes vertically in columns.

I wish you success with your math teaching!

Maria Miller

Helpful Resources on the Internet

Use these free online resources to supplement the "bookwork" as you see fit.

Clock - Teaching Tool
An interactive clock that you can use to demonstrate telling time or elapsed time.
http://mrcrammond.com/games/clock.swf

Interactive Teaching Clock
Play around with the controls—you can do all kinds of demonstrations with this clock!
http://www.visnos.com/demos/clock

That Quiz: Time
Online quizzes for all time-related topics: reading the clock, time passed, adding/subtracting with time, conversion of time units, and time zone practice. The quizzes have many levels, can be timed or not, and include lots of options for customization. Easy to use and set up.
http://www.thatquiz.org/tq-g/math/time

Difference Between Two Times
Work out the difference between two given times. A time line may help.
http://www.bgfl.org/bgfl/custom/resources_ftp/client_ftp/ks2/maths/timetables/index.htm

Elapsed Time
Click "New Time". Then click the buttons that advance the time on the clock, until the time matches the "End" time. Choose difficulty levels 1 and 2 for this grade level.
http://www.shodor.org/interactivate/activities/ElapsedTime/

Time Difference
Demonstrate elapsed time. Set the start and end times on the two clocks. Then, click the step-counters ("1 hr", "5 min", and "1 min") and the first clock will go ahead and keep track of elapsed time.
http://mathszone.webspace.virginmedia.com/mw/time/Time%20Difference.swf

Elapsed Time Worksheets
Generate printable worksheets for elapsed time. You can practice the elapsed time, finding the starting time, or finding the ending time.
http://www.mathnook.com/elapsedtimegen.html

Find the Start Time
Word problems about starting times with multiple-choice answers. Choose "full screen", then "Find the start time". Next, choose option 4 or 5.
http://mathsframe.co.uk/en/resources/resource/119/find_the_start_time

Time for Crime — Elapsed Time Mystery
A single mystery problem which can be solved by thinking of the elapsed time: who is the thief?
http://teacher.scholastic.com/maven/timefor/index.htm

ThatQuiz — Elapsed time
A ten-question quiz on Elapsed Time.
http://www.thatquiz.org/tq-g/?-j4-l4-p0

Adding Time Word Problems
Read the time and then answer a word problem involving adding a given time.
http://mathsframe.co.uk/en/resources/resource/118/adding_time_word_problems

Interactivate: Elapsed Time
Practice calculating elapsed time with analog or digital clocks.
http://www.shodor.org/interactivate/activities/ElapsedTime/

RoboClock 3: Elapsed Time
How much time has passed in hours and minutes? Help RoboClock solve the problems.
http://www.primarygames.com/math/roboclock3/

Time Slice
Solve word problems involving time and elapsed time.
http://www.mathslice.com/time-slice.html

How Many Hours Pass?

The chart below shows the whole hours in one 24-hour period = one night + one day.

12 – 1 – 2 – 3 – 4 – 5 – 6 – 7 – 8 – 9 – 10 – 11 – 12 – 1 – 2 – 3 – 4 – 5 – 6 – 7 – 8 – 9 – 10 – 11 – 12

Midnight **AM** **Noon** **PM** **Midnight**

From midnight to noon we call the hours "AM". This comes from *Ante Meridiem* (Latin), and means *before noon.* From noon to midnight we call the hours "PM", which comes from *Post Meridiem* (Latin), and means *after noon.*

How many hours is it from 6 AM to 11 AM?

You could use the chart, and count. But since both hours are AM, you can use subtraction to find the difference: 11 − 6 = 5 hours.

How many hours is it from 3 AM to 3 PM?

Now you cannot use subtraction because the answer clearly is not zero hours. Since the number is the same (3), it means the hour hand travels through the entire clock face, starting at 3 and ending at 3. The difference is 12 hours.

How many hours is it from 8 AM to 3 PM?

One of the times is AM, and the other is PM, so you cannot subtract them. Instead, do it in two parts:

1) How many hours from 8 AM till noon? It is four hours.
2) How many hours from noon till 3 PM? It is three hours.

All totaled, there are 7 hours from 8 AM to 3 PM.

1. How many hours is it?

from	5 AM	7 AM	9 AM	11 AM	10 AM
to	12 noon	1 PM	4 PM	11 PM	7 PM
hours					

2. How long is the school day, if it starts and ends at given times?

Start:	8 AM	8 AM	9 AM	10 AM	8 AM
End:	12 noon	1 PM	3 PM	3 PM	2 PM
hours:					

3. How many hours is it till midnight?

from	4 PM	7 PM	12 noon	9 AM	7 AM
to	12 midnight	12 midnight	12 midnight	12 midnight	12 midnight
hours					

4. How many hours does Matthew sleep if he goes to bed and gets up at given times?

Go to bed	9 PM	8 PM	9 PM	11 PM	12 midnight
Get up	6 AM	7 AM	5 AM	9 AM	9 AM
Sleep hours					

5. **a.** How many hours is your school day usually?

b. How many hours do you usually sleep?

6. **a.** Dad's workday starts at 8:00 in the morning, and ends at 5 PM.
How many hours is Dad at work?

b. Mary's school day starts at 9 AM and ends at 2 PM. How long is her school day?

c. The airplane took off at 10 AM and landed at 1 PM. Then it took off again at 2 PM and landed at 6 PM. How many hours was the airplane in the air?

7. **a.** How many hours are there in one day-night period?

b. How many hours are there in two day-night periods?

8. **a.** The turkey needs to cook three hours in the oven to be ready at 7 PM.
When should it be put into the oven?

b. It takes two hours to mow the lawn. Jim wants to be done at 1 PM.
When should he start mowing?

c. Mom needs seven hours of sleep tonight. She wants to wake up at 6 AM.
When should she go to bed?

How Many Minutes Pass?

1. How many minutes does the minute hand “cover,” or “pass through,” on the clock? Count by fives. Also, show the time passing using your practice clock.

	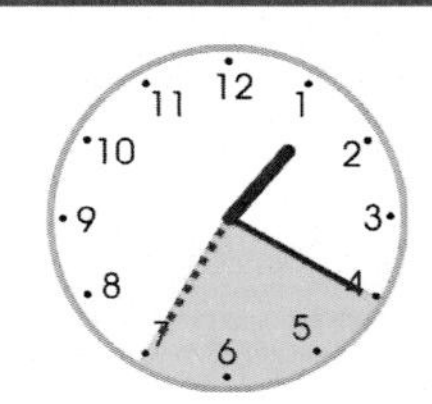		
a. From 10:00 till 10:15	**b.** From 1:20 till 1:35	**c.** From 5:50 till 6:10	**d.** From 2:05 till 2:40
______ minutes	______ minutes	______ minutes	______ minutes

2. Make your practice clock show the starting time. Then, move the minute hand till the ending time. How many minutes pass? Count by fives.

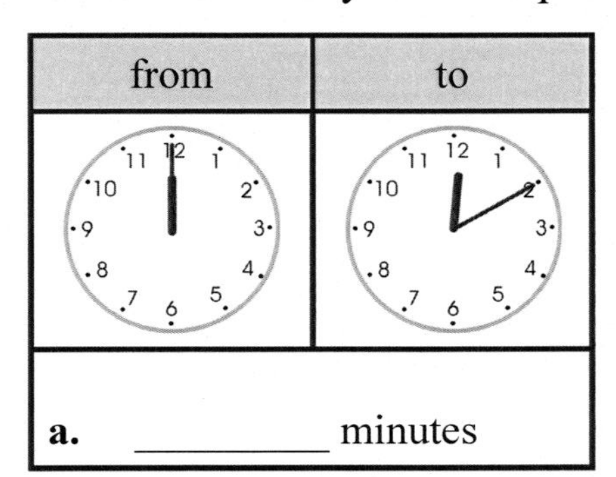

a. __________ minutes

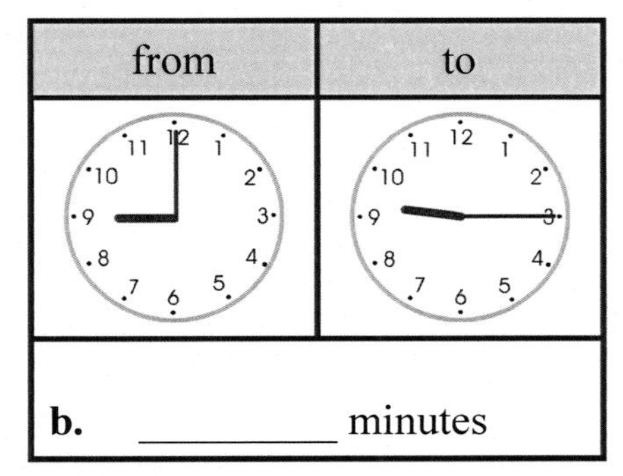

b. __________ minutes

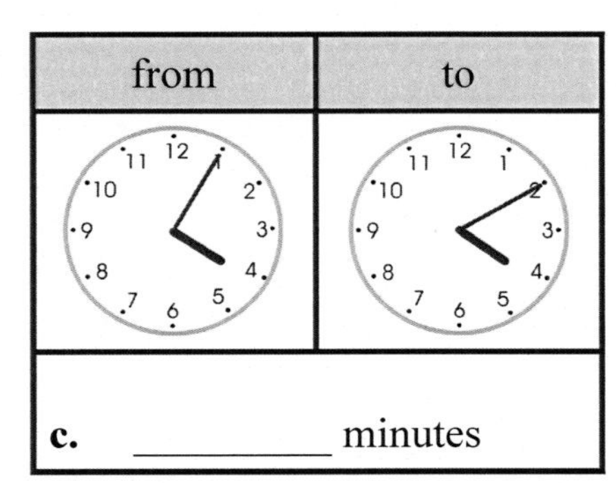

c. __________ minutes

from	to

d. __________ minutes

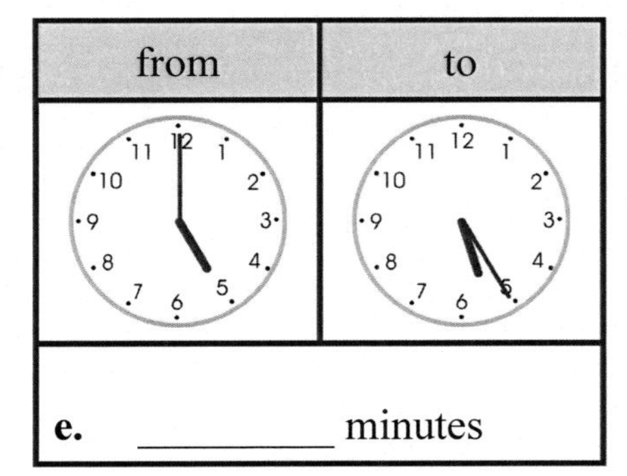

e. __________ minutes

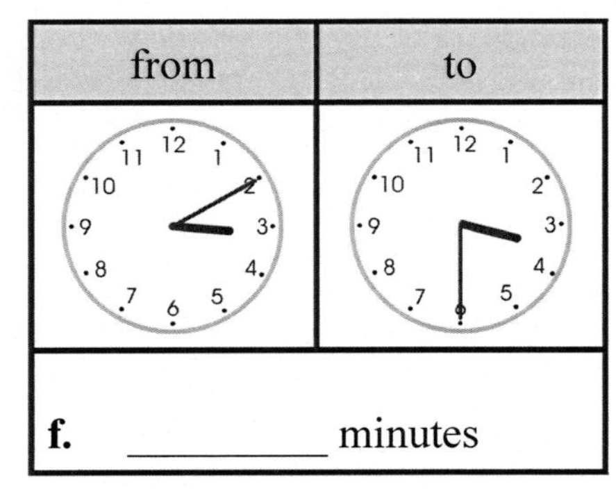

f. __________ minutes

3. The clock shows the time now. Write the later times. Use your practice clock to help, or imagine the minute hand moving ahead.

- 5 minutes later _____ : ______
- 10 minutes later _____ : ______
- 20 minutes later _____ : ______
- 25 minutes later _____ : ______

4. The class ends at 1:00. How many minutes of class are left at these times?

from	to
	1:00
a. _______ minutes	

from	to
	1:00
b. _______ minutes	

from	to
	1:00
c. _______ minutes	

from	to
	1:00
d. _______ minutes	

5. The clock shows the time now. Write the later times. Use your practice clock to help, or imagine the minute hand moving ahead.

NOW:

a.

- 5 minutes later _____ : ______
- 10 minutes later _____ : ______
- 20 minutes later _____ : ______
- 25 minutes later _____ : ______

NOW:

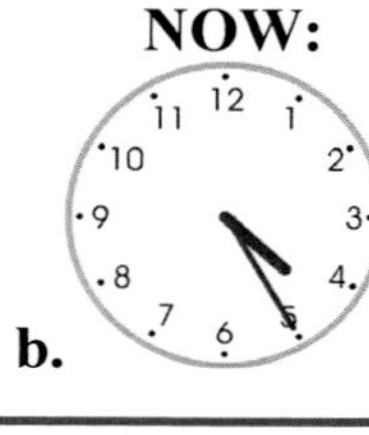

b.

- 10 minutes later _____ : ______
- 15 minutes later _____ : ______
- 30 minutes later _____ : ______
- 35 minutes later _____ : ______

NOW:

c.

- 5 minutes later _____ : ______
- 10 minutes later _____ : ______
- 15 minutes later _____ : ______
- 25 minutes later _____ : ______

6. How many minutes till 2:30? Till 8:00? Use your practice clock to help.

a.

It is 2:00. → ______ minutes till

It is 2:10. → ______ minutes till

It is 2:20. → ______ minutes till

It is 2:25. → minutes till

2:30

b.

It is 7:30. → ______ minutes till

It is 7:35. → ______ minutes till

It is 7:45. → ______ minutes till

It is 7:50. → minutes till

8:00

If two different times have the same hour, you can <u>subtract</u> to find how many minutes pass.
Example. How many minutes pass from 11:10 to 11:20? Since the hours are both 11, just look at the minute-amounts (10 and 20), and subtract them. 20 − 10 = 10. So, 10 minutes pass.

7. How many minutes pass? Subtract (or figure out the difference).

from	1:25	2:00	3:05	7:30	5:10
to	1:55	2:15	3:25	7:50	5:50
minutes	<u>30 minutes</u>				

from	2:00	7:05	8:25	6:40	11:15
to	2:35	7:35	8:50	6:55	11:40
minutes					

8. **a.** The bus trip started at 4:10, and ended at 4:30. How many minutes did it take?

b. Joshua started math homework at 4:40, and ended at 5:05.
How much time did he spend?

c. Music class starts at 10:15, and ends at 10:45. How long is the class?

From 2:45 to 3:45 the minute hand makes <u>one full circle</u>, starting and ending in the same position (45 minutes). Therefore, from 2:45 to 3:45 is <u>one hour</u>.

How much time passes from 7:20 to 10:20? Since the minute-amounts are the same (20 minutes), the minute hand has made some full rounds around the clock. How many full rounds? Just look at the difference between the hour-amounts: 7 and 10. This means three hours pass from 7:20 to 10:20.

9. The minute hand makes full rounds around on the clock. How many whole hours pass?

from	10:30	10:30	1:40	5:45	3:20 AM
to	11:30	12:30	4:40	11:45	12:20 PM
full rounds or hours					

Elapsed Time

How many minutes is it till the next whole hour?

It is 4:38. The minute hand needs to go 2 minutes till the 40-minute point (number 8), and then 20 more minutes till the next whole hour. So it is 22 minutes till 5 o'clock.

Or, you can subtract 38 minutes from 60 minutes: $60 - 38 = 22$. Remember, a complete hour is 60 minutes.

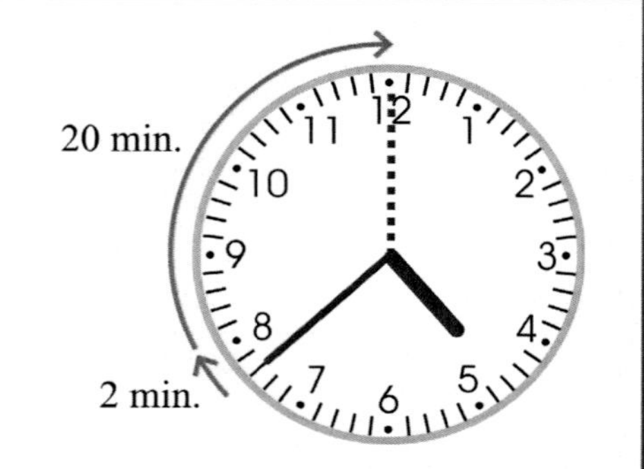

It is 2:34. How many minutes is it till 2:50?

The hour is the same (2 hours) in both times, you can simply subtract the minutes: $50 - 34 = 16$ minutes.

Or, add up from 34 till 50:
$34 + \mathbf{6} = 40$
$40 + \mathbf{10} = 50$.
You added 16 minutes.

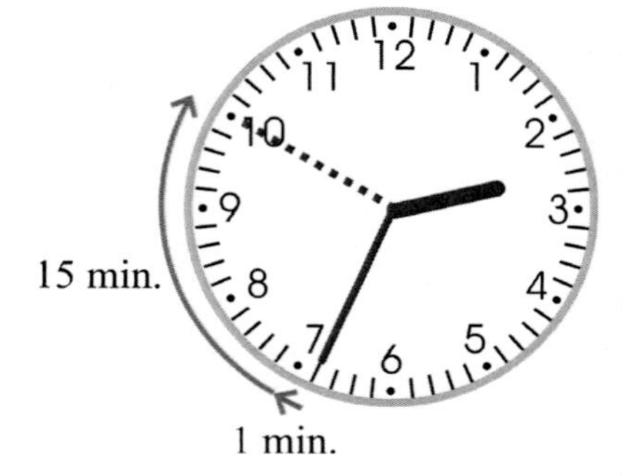

Or, imagine the minute hand moving on the clock face: it moves 1 minute, and then another 15 minutes — a total of 16 minutes.

1. How many minutes is it till the next whole hour?

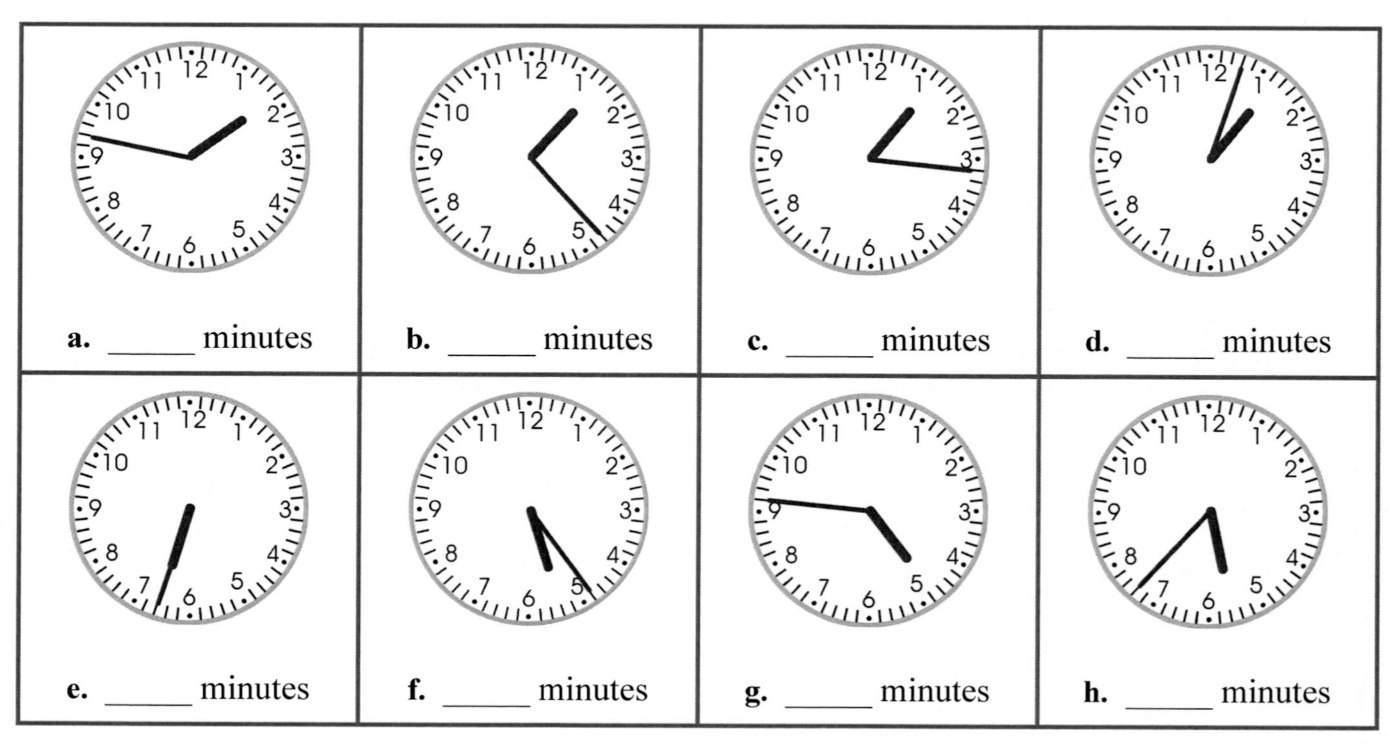

2. How many minutes is it from the time on the clock face till the given time?

till 12:40 **a.** _____ minutes	till 7:30 **b.** _____ minutes	till 10:45 **c.** _____ minutes	till 3:58 **d.** _____ minutes
till 1:00 **e.** _____ minutes	till 5:55 **f.** _____ minutes	till 12:50 **g.** _____ minutes	till 4:55 **h.** _____ minutes

3. How many minutes is it?

a. From 5:06 till 5:28	**b.** From 2:05 till 2:54	**c.** From 3:12 till 3:47
d. From 12:11 till 12:55	**e.** From 7:27 till 7:48	**f.** From 9:06 till 10:00

4. **a.** The pie needs to bake half an hour. Marsha's clock showed 4:22 when she put it into the oven. When should she take it out?

b. Juan notices that, "In 14 minutes class will end." If the class ends at 2 PM, what time is it now?

c. The sun rises at 5:49 AM. Marge wants to wake up 15 minutes before that. When should she wake up?

d. Edward was 8 minutes late to math class, and arrived at 1:53 PM. When did the class start?

More about Elapsed Time

How many minutes is it from 1:47 to 2:10?

Notice that the hour changes from 1 to 2. We need to calculate this carefully, but it is easy when you calculate it in two parts:

From 1:47 to 2:00 is 13 minutes. From 2:00 to 2:10 is 10 minutes. So the total is 23 minutes.

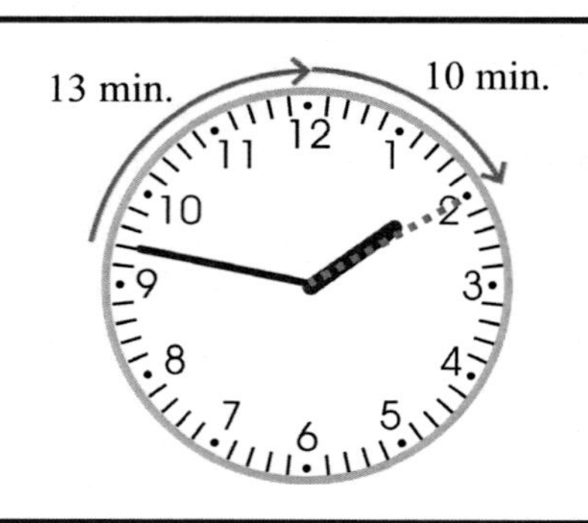

How many minutes is it from 4:28 to 5:15?

Again, the hour changes, so we figure it in two parts:
From 4:28 to 5:00 is 32 minutes. From 5 to 5:15 is 15 minutes. The total is: 32 + 15 = 47 minutes.

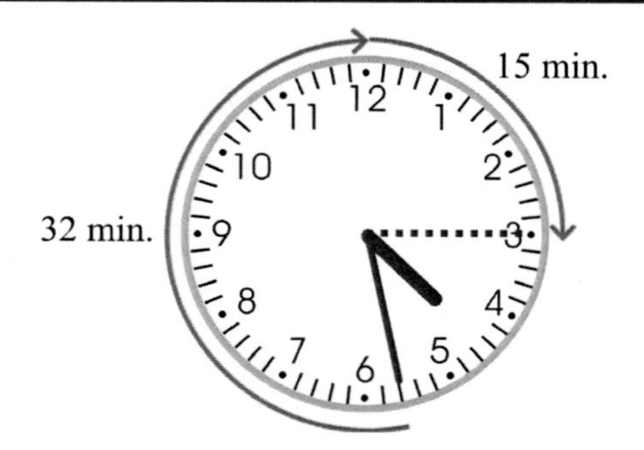

1. How many minutes is it from the time on the clock face until the given time?

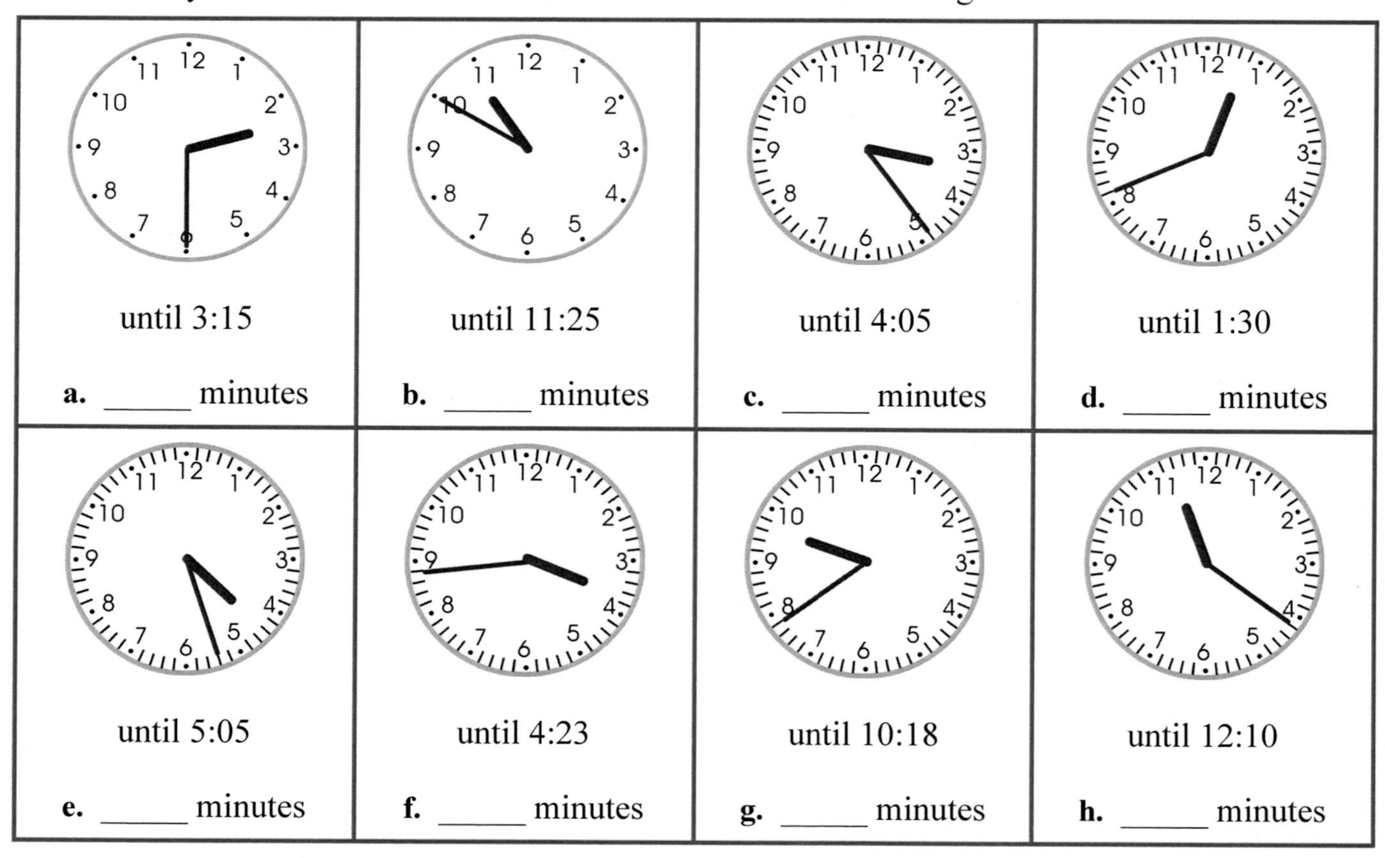

until 3:15 **a.** _____ minutes	until 11:25 **b.** _____ minutes	until 4:05 **c.** _____ minutes	until 1:30 **d.** _____ minutes
until 5:05 **e.** _____ minutes	until 4:23 **f.** _____ minutes	until 10:18 **g.** _____ minutes	until 12:10 **h.** _____ minutes

How much time passes from 5:38 to 8:38?

The minutes are the same (:38), so the minute hand has made some full rounds—full hours—and ended up back in the same place. So you need to look only at the difference in the hours: From 5 to 8 is 3 hours. So three hours have passed.

How much time passes from 11:30 to 4:30?

Once again, the minute hand has made several full rounds. From 11 to 4 is five hours. Check this by turning the hands on your practice clock!

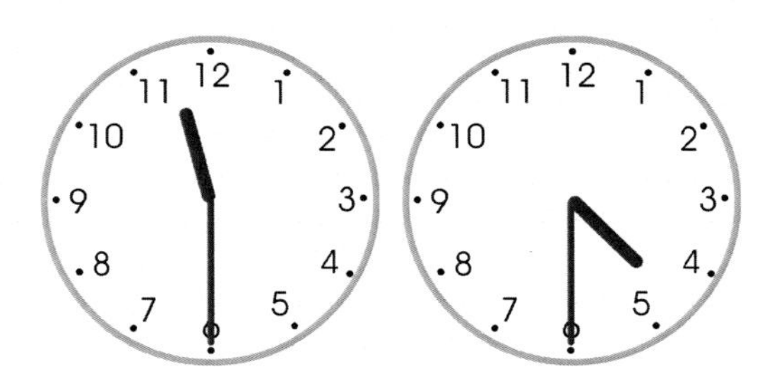

You can also figure the passed time in parts:
(1) From 11:30 to 12:00 is half an hour.
(2) From 12:00 to 4:00 is four more hours.
(3) From 4:00 to 4:30 is another half an hour.
The total is four hours plus two half hours,
but two half hours makes another whole hour.
So again, the total is five hours.

For combinations of whole hours and half hours that go by—calculate in parts.

a. How much time passes from 5:00 to 7:30?

From 5:00 to 7:00 is two whole hours.
From 7:00 to 7:30 is half an hour.
Total: two and a half hours.

b. How much time passes from 11:30 AM to 5:00 PM?

From 11:30 AM to 12:00 noon is half an hour.
From 12:00 to 5:00 PM is five hours.
Total: five and a half hours.

2. How much time passes during these intervals?

a. From 2:06 to 10:06	**b.** From 8:25 to 12:25	**c.** From 3:30 to 6:00
d. From 7:30 AM to 1:30 PM	**e.** From 10:00 AM to 3:30 PM	**f.** From 9:49 to 1:49
g. From 5 AM to 5 PM	**h.** From 11 PM to 12 noon	**i.** From 6 AM to 4 PM

3. How much time passes? Figure it out in parts.

<table>
<tr>
<td>

a. From 1:40 to 2:30

From 1:40 to 2:00 = ______ minutes

From 2:00 to 2:30 = ______ minutes

Total = ______ minutes

</td>
<td>

b. From 7:30 AM to 3:10 PM

From 7:30 to 8:00 = ____ h ____ min.

From 8:00 to 12:00 = ____ h ____ min.

From 12:00 to 3:00 = ____ h ____ min.

From 3:00 to 3:10 = ____ h ____ min.

Total = ____ h ____ min.

</td>
</tr>
<tr>
<td>

c. From 2:35 to 8:15

</td>
<td>

d. From 6:40 to 4:15

</td>
</tr>
</table>

4. Write the ending time or the starting time. Imagine turning the minute hand on a clock, or use your practice clock.

a. 6:15 → _______ 40 minutes	**b.** 2:03 → _______ 25 minutes	**c.** 11:30 → _______ 35 minutes
d. _______ → 5:50 35 minutes	**e.** _______ → 7:00 45 minutes	**f.** _______ → 12:10 20 minutes

5. Solve the problems. Imagine turning the minute hand on a clock.

a. The science class starts at 10:55 and ends 50 minutes later.
What time does it end?

b. An airplane took off at 3:35 PM and landed at 7:10 PM.
How long was the flight?

c. Eddie played with building blocks for 2 hours 40 minutes, starting at 3:30.
What time did he stop?

d. Jane practiced the piano for 45 minutes, starting at 2:30. What time was she done?

6. Write what time it is in the story as it goes along.

It was 4:20, and Mr. Toad was sitting in his favorite chair at home, excited because in fifteen minutes he was going to go out with his friend, Mr. Rat, to do something special. Soon he heard the doorbell ring. Mr. Rat was at the door and it was time to go. (Time: ________)

It was a 20-minute walk to Mr. Mole's house, and the two friends were planning not to be late. After walking for 10 minutes (Time: ________), they heard a familiar voice from behind them, "Hey, where are you going so fast?"

It was Mr. Duck, who was wandering around without any particular place to go in mind. The three friends chatted for five minutes (Time: ________), and then they decided to go together to surprise Mr. Mole.

Soon they arrived at Mr. Mole's house (Time: ________), and knocked on the door. Mr. Mole had no idea they were coming, and he was very surprised and happy to see his friends. After chatting for ten minutes (Time: ________), they headed for the kitchen to sip some cranberry juice.

7. Here is part of a television schedule. Answer the questions about the programs.

Channel 1	**Channel 2**	**Channel 3**
4:30 Nature film: Whales 5:30 Children's Story Time 6:05 Early News 6:35 Shopping Spree Show 7:05 The Week in Politics	4:30 Cooking Class 5:05 Kids TV 5:55 Quick News 6:20 Nature Film: The Antarctic 7:25 Current Trends	4:45 Afternoon Bits 5:15 Nature film: Bees and Honey 6:20 Flash News 6:40 The Silly Faces Show 7:20 Arnold's Kitchen

a. Each of the three channels has a nature film.
List here how long each one of them lasts.

b. Which is the longer program, "Children's Story Time" or "Kids TV"?
How many minutes longer?

c. Which channel has the *longest* news program?

Which one has the *shortest* news program?

What is the difference between the two programs in minutes?

d. Megan changed channels like this:

From 4:30 to 5:15 Channel 1
From 5:15 to 6:20 Channel 3
From 6:20 to 7:25 Channel 2

Which programs did Megan watch (either totally or partially)?

Elapsed Time or How Much Time Passes

<table>
<tr><td>You can find any time difference—elapsed time—by starting from the earlier time and "adding up" the elapsing time until the latter time. Imagine turning the hand of a clock from the starting time on, and keeping track of how much time passes.

How long was an airplane flight if the plane took off at 12:45 p.m. and landed at 5:10 p.m. ?</td><td>From 12:45 till 1:00 | 15 min
From 1 till 5 | 4 h
From 5 till 5:10 | 10 min
Total | 4 h 25 min</td></tr>
<tr><td>Time differences can also be found by <u>subtracting</u>. Subtract the hours and minutes separately in their own columns.

How much time passes between 2:10 a.m. and 8:43 a.m.?</td><td>8 h 43 m
− 2 h 10 m
6 h 33 m</td></tr>
<tr><td>How much time passes between 4:46 p.m. and 7:13 p.m.?

You cannot subtract 46 minutes from 13 minutes. You need to <u>regroup</u> 1 hour as 60 minutes. So add 60 to the minutes count. Do NOT regroup 1 hour as 10 or 100 minutes!</td><td>6 h 73 m
<s>7 h</s> <s>13 m</s>
− 4 h 46 m
2 h 27 m</td></tr>
<tr><td>How much time passes between 9:42 p.m. and 2:45 a.m.?

Here the p.m. changes to a.m. It is safer to figure this in parts: from 9:42 p.m. to midnight, and from midnight to 2:45 a.m.

If you subtracted the two numbers, you would get the time difference the other way around: From 2:45 to 9:42, which is not the right answer. (Of course, knowing that, you could figure out the other by subtracting the answer from 12 hours.)</td><td>9:42 p.m .... 10 p.m. = 18 min
10 p.m .... midnight = 2 hours
Midnight ... 2:45 a.m. = 2 h 45 min

18 m
2 h 0 m
+ 2 h 45 m
4 h 63 m = 5 h 3 m</td></tr>
</table>

1. How much time passes? Solve mentally.

a. From 12:30 p.m. till 2 p.m. _____ h _______ min	**b.** From 4:35 p.m. till 6:15 p.m. _____ h _______ min	**c.** From 5:19 a.m. till noon _____ h _______ min
d. From 9:30 a.m. till 2:10 p.m. _____ h _______ min	**e.** From 7.58 p.m. till midnight _____ h _______ min	**f.** From 11:05 p.m. till 6:35 a.m. _____ h _______ min

2. How much time passes? Use subtraction.

a. From 4:53 p.m. till 8:26 p.m. 8 h 26 m − 4 h 53 m	**b.** From 6:37 p.m. till 9:03 p.m.	**c.** From 2:45 a.m. till 8:14 a.m.

3. How much time passes? Do it in two parts, since one time has a.m. and the other has p.m.

a. From 8:27 p.m. till 2:12 a.m.	**b.** From 9 a.m. till 5:16 p.m.	**c.** From 10:48 a.m. till 8:26 p.m.

4. How much time passes? Now we use the 24-hour clock.

a. From 8:27 till 13:45	**b.** From 6:30 till 17:10	**c.** From 9:45 till 23:25

5. Make a schedule for a doctor. He assigns 30 minutes for each patient, and after three patients, he has a 20-minute break. Use the 24-hour clock.

	Time
Patient 1	8:00 - 8:30
Patient 2	
Patient 3	
break	
Patient 4	
Patient 5	
Patient 6	
break	

	Time
Patient 7	
Patient 8	
Patient 9	
break	
Patient 10	
Patient 11	
Patient 12	

6. Make a class schedule. Each class is 50 minutes, with 5 minutes between them. The lunch break is 40 minutes.

Class	**Time**
Social Studies	8:00 -
Math	
Science	
English	

Class	**Time**
Lunch	
History	
P.E.	

7. If you would like, make a schedule for yourself. Include things such as school work, chores, meals, and play time.

When will it end?

The meeting starts at 2:30 p.m. and lasts for 1 hour 15 minutes.

Simply add the hours to the clock-time hours, and minutes to the clock-time minutes:

2 hours + 1 hour = 3 hours.
30 minutes + 15 minutes = 45 minutes.
Answer: The meeting ends at 3:45 p.m.

Jake started playing at 3:35 p.m. and played for 45 minutes.

You can still add like you did above and get 3 hours 80 minutes, but 80 minutes is more than one hour! We need to see the 80 minutes as 60 + 20, where 60 minutes makes one hour. Therefore **the final answer is** 4 hours and 20 minutes, or 4:20 p.m.

The other way is to add the starting time and the elapsing time.

If it started raining at 10:53 and it rained for 4 hours and 40 minutes, when did the rain end?

Add the minutes and the hours separately. Note the minutes go over 60, so we need to change the 93 minutes to 1 hour and 33 minutes. The final answer is 15:33 or 3:33 p.m.

$$\begin{array}{rrr} & 10\text{ h} & 53\text{ m} \\ + & 4\text{ h} & 40\text{ m} \\ \hline & 14\text{ h} & 93\text{ m} \\ = & 15\text{ h} & 33\text{ m} \end{array}$$

8. When will it end?

a. Guests came at 3:40 p.m. and stayed for two hours and 30 minutes.

b. Making pizza takes one hour and 40 minutes. Mom starts at 13:45.

c. The pool opens at 8 a.m. and is open for 10 and a half hours. When does it close?

d. Jen's exam lasted for two and a half hours, starting at 8:45.

e. The airplane takes off at 18:08 and flies for three hours and 55 minutes.

f. The food went into the oven at 5:47 p.m. and baked for 35 minutes.

g. Workers in a factory work in three shifts.
How long is each shift?

Shift 1	6:00 a.m. - 2:30 p.m.
Shift 2	2:00 p.m. - 10:00 p.m.
Shift 3	9:30 p.m. - 6:30 a.m.

How many minutes is the overlap between two shifts?

When did it start?

One more possible problem is that you know when something ends and how long it lasted.

The airplane landed at 4:30 p.m. The flight lasted for 3 hours and 40 minutes. When did the plane take off?

You need to go backwards from the ending time. Start at 4:30 and let the minute hand travel in your mind backwards 3 full rounds, and then 40 minutes. Where do you end up?

Alternatively, subtract in columns. You will again need to regroup an hour as 60 minutes. The answer 50 minutes would mean the clock time 12:50 p.m.

	3h	90 m
	~~4 h~~	~~30 m~~
−	3 h	40 m
		50 m

Mental math is always good!

A 55-minute class ended at 21:10. When did it start?

If it had lasted 1 hour, it would have started at 20:10.
But it was 5 minutes shorter and therefore started 5 minutes later, or at 20:15.

A TV show lasted 1 h and 35 min, ending at 11:20 p.m. When did it start?

Subtract in columns or think it through mentally. Again you need to regroup 1 hour as 60 minutes. It started at 9:45 p.m.

	10 h	80 m
	~~11 h~~	~~20 m~~
−	1 h	35 m
	9 h	45 m

9. Find the starting time.

a. From _____ : _______ p.m. till 2:00 p.m. is 40 minutes.

b. From _____ : _______ p.m. till 8:12 p.m. is 30 minutes.

c. From _____ : _______ a.m. till 4:15 a.m. is 1 hour 30 minutes.

d. From _____ : _______ p.m. till 7:34 p.m. is 4 hours10 minutes.

e. From _____ : _______ a.m. till 5:00 p.m. is 6 hours 20 minutes.

f. From _____ : _______ a.m. till 6:54 a.m. is 5 hours 32 minutes.

g. From _____ : _______ p.m. till 15:30 p.m. is 45 minutes.

h. From _____ : _______ p.m. till 16:30 p.m. is 2 hours 40 minutes.

10. Solve.

a. The Johnson family arrived in the city at 10:30 after a 3-hour, 15-minute car ride. When did they leave home?

b. When should the family leave the city to make it home by 20:00 (assuming the driving time back home is the same)?

c. Shannon runs through a path in the woods, and times himself. Complete the chart with the amount of time he spent running each day.

	Mo	**Wd**	**Th**	**Fr**	**Sa**
Start: End:	17:15 18:20	17:03 18:05	17:05 18:12	17:45 18:39	17:12 18:15
Running time:					

d. Find Shannon's total running time during the week.

e. Gordon works from 8:30 until 17:15 each day. He has a 30-minute lunch break, and two 15-minute "coffee" breaks. How many hours/minutes does he actually work?

f. The air conditioner is kept running from 7:30 a.m. until 9 p.m. How many hours does it run in a week?

g. An airplane is scheduled to take off at 3:40 p.m. and land at 5:10 p.m. The flight is delayed so that it leaves at 3:55 p.m. instead. When will it land?

Review

1. How much time passes?

   a. From 11:15 p.m. till 6:07 a.m.

   b. From 10:55 till 21:35.

2. A flight that lasts 4 hours 20 minutes took off at 1:50 p.m.
   When does it land?

3. How many minutes is it from the time on the clock face until the given time?

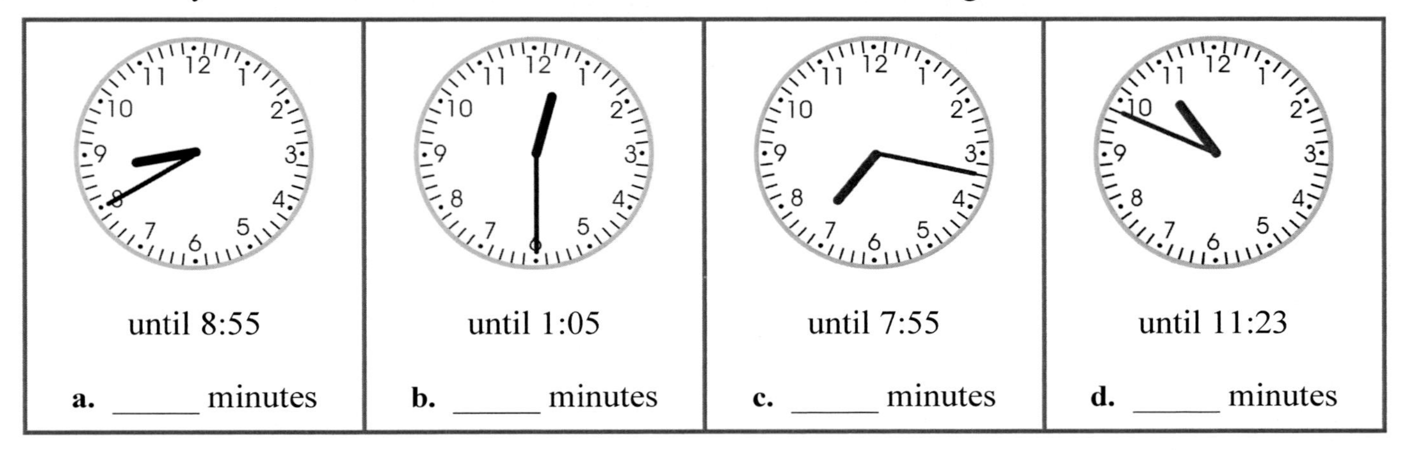

until 8:55	until 1:05	until 7:55	until 11:23
a. ______ minutes	**b.** ______ minutes	**c.** ______ minutes	**d.** ______ minutes

4. How much time passes between the two times given?

a. from 4:08 until 10:08	**b.** from 3 AM until 5 PM
c. from 8:23 until 8:41	**d.** from 3:37 until 4:17

5. The music class starts at 1:45 and ends 50 minutes later.
   At what time does it end?

6. The train left at 11:10 and arrived at 12:20 PM.
   How long was the trip?

Elapsed Time Answer Key

How Many Hours Pass?, p. 9

1.

from	5 AM	7 AM	9 AM	11 AM	10 AM
to	12 noon	1 PM	4 PM	11 PM	7 PM
hours	7	6	7	12	9

2.

Start:	8 AM	8 AM	9 AM	10 AM	8 AM
End:	12 noon	1 PM	3 PM	3 PM	2 PM
hours:	4	5	6	5	6

3.

from	4 PM	7 PM	12 noon	9 AM	7 AM
to	12 midnight	12 midnight	12 midnight	12 midnight	12 midnight
hours	8	5	12	15	17

4.

Go to bed	9 PM	8 PM	9 PM	11 PM	12 midnight
Get up	6 AM	7 AM	5 AM	9 AM	9 AM
Sleep hours	9	11	8	10	9

5. Please check the student's answer. Answers will vary.

6. a. Dad is at work for nine hours.
   b. Mary's school day is five hours long.
   c. The airplane was in the air for seven hours.

7. a. There are 24 hours in a day-night period.
   b. There are 48 hours in 2 day-night periods.

8. a. The turkey should be put into the oven at 4 p.m.
   b. He should start mowing at 11 a.m.
   c. Mom should go to bed at 11 p.m.

How Many Minutes Pass?, p. 11

1. a. 15 minutes b. 15 minutes c. 20 minutes d. 35 minutes

2. a. 10 minutes b. 15 minutes c. 5 minutes d. 10 minutes e. 25 minutes f. 20 minutes

3.

5 minutes later 6:50
10 minutes later 6:55
20 minutes later 7:05
25 minutes later 7:10

4. a. 45 minutes b. 35 minutes c. 20 minutes d. 15 minutes

5.

a.

5 minutes later 8:25
10 minutes later 8:30
20 minutes later 8:40
25 minutes later 8:45

b.

10 minutes later 4:35
15 minutes later 4:40
30 minutes later 4:55
35 minutes later 5:00

c.

5 minutes later 10:50
10 minutes later 10:55
15 minutes later 11:00
25 minutes later 11:10

6.

a.		b.	
It is 2:00. → 30 minutes till It is 2:10. → 20 minutes till It is 2:20. → 10 minutes till It is 2:25. → 5 minutes till	2:30	It is 7:30. → 30 minutes till It is 7:35. → 25 minutes till It is 7:45. → 15 minutes till It is 7:50. → 10 minutes till	8:00

7.

from	1:25	2:00	3:05	7:30	5:10
to	1:55	2:15	3:25	7:50	5:50
minutes	30 minutes	15 minutes	20 minutes	20 minutes	40 minutes

from	2:00	7:05	8:25	6:40	11:15
to	2:35	7:35	8:50	6:55	11:40
minutes	35 minutes	30 minutes	25 minutes	15 minutes	25 minutes

8. a. It took 20 minutes for the bus trip.
   b. Joshua spent 25 minutes doing math.
   c. Music class is 30 minutes long.

9.

from	10:30	10:30	1:40	5:45	3:20 AM
to	11:30	12:30	4:40	11:45	12:20 PM
full rounds or hours	1 hour	2 hours	3 hours	6 hours	9 hours

Elapsed Time, p. 14

1. a. 13 minutes b. 37 minutes c. 44 minutes d. 57 minutes
   e. 27 minutes f. 36 minutes g. 14 minutes h. 23 minutes

2. a. 19 minutes b. 13 minutes c. 36 minutes d. 27 minutes
   e. 19 minutes f. 31 minutes g. 49 minutes h. 28 minutes

3. a. 22 minutes b. 49 minutes c. 35 minutes d. 44 minutes e. 21 minutes f. 54 minutes

4. a. At 4:52 she should take it out. b. It is 1:46 PM.
   c. She should wake up at 5:34 AM. d. The class started at 1:45 PM.

More about Elapsed Time, p. 16

1. a. 45 minutes b. 35 minutes c. 41 minutes d. 49 minutes
   e. 38 minutes f. 39 minutes g. 39 minutes h. 49 minutes

2. a. 8 hours b. 4 hours c. 2 1/2 hours d. 6 hours e. 5 1/2 hours
   f. 4 hours g. 12 hours h. 13 hours i. 10 hours

3. a. Total 50 minutes b. Total 7 hours and 40 minutes
   c. Total 5 hours and 40 minutes d. Total 9 hours and 35 minutes

4. a. 6:55 b. 2:28 c. 12:05 d. 5:15 e. 6:15 f. 11:50

5. a. It ends at 11:45. b. The flight was 3 hours and 35 minutes long.
   c. Eddie stopped at 6:10 d. Jane stopped at 3:15.

6. First paragraph (time 4:35), second paragraph (time 4:45), third paragraph (time 4:50), fourth paragraph (time 5:00) and (time 5:10).

7. a. Channel 1: 1 hour, Channel 2: 1 hour and 5 minutes, Channel 3: 1 hour and 5 minutes.
   b. Kids TV Time is 15 minutes longer.
   c. Channel 1 has the longest news, Channel 3 has the shortest news, and the difference in minutes is 10 minutes.
   d. Channel 1, Megan watched part of Nature film: Whales, Channel 3, she watched all of Nature film: Bees and Honey, and on Channel 2, she watched all of Nature film: The Antarctic.

Elapsed Time or How Much Time Passes, p. 20

1. a. 1 hr 30 min b. 1 hr 40 min c. 6 hr 41 min d. 4 hr 40 min e. 4 hr 2 min f. 7 hr 30 min

2. a. 3 hr 33 min b. 2 hr 26 min c. 5 hr 29 min

3. a. 5 hr 45 min b. 8 hr 16 min c. 9 hr 38 min

4. a. 5 hr 18 min b. 10 hr 40 min c. 13 hr 40 min

Elapsed Time or How Much Time Passes, cont.

5.

	Time
Patient 1	8:00 - 8:30
Patient 2	8:30 - 9:00
Patient 3	9:00 - 9:30
break	9:30 - 9:50
Patient 4	9:50 - 10:20
Patient 5	10:20 - 10:50
Patient 6	10:50 - 11:20
break	11:20 - 11:40

	Time
Patient 7	11:40 - 12:10
Patient 8	12:10 - 12:40
Patient 9	12:40 - 13:10
break	13:10 - 13:30
Patient 10	13:30 - 14:00
Patient 11	14:00 - 14:30
Patient 12	14:30 - 15:00

6.

Class	Time
Social Studies	8:00 - 8:50
Math	8:55 - 9:45
Science	9:50 - 10:40
English	10:45 - 11:35

Class	Time
Lunch	11:35 - 12:15
History	12:15 - 1:05
P.E.	1:10 - 2:00

7. Answers will vary. Please check the student's work.

8. a. 6:10 b. 15:25 c. 18:30 d. 11:15 e. 22:03 f. 6:22
g. Shift 1: 8 hours 30 min; Shift 2: 8 hours; Shift 3: 9 hours. Each shift overlaps the next by 30 minutes.

9. a. 1:20 p.m. b. 7:42 p.m. c. 2:45 p.m. d. 3:24 p.m.
e. 10:40 a.m. f. 1:22 a.m. g. 14:45 p.m. h. 13:50 p.m.

10. a. They left home at 7:15.
b. They should leave the city by 16:45.

c.	Mo	Wd	Th	Fr	Sa
Start: End:	17:15 18:20	17:03 18:05	17:05 18:12	17:45 18:39	17:12 18:15
Running time:	1 h 5 min	1 h 2 min	1 h 7 min	54 min	1 h 3 min

d. His total running time was 5 hours and 11 minutes.
e. From 8:30 until 12:00 is 3 hours 30 min. Then, from 12:00 until 17:15 is 5 hours 15 min. 3 hours 30 min + 5 hours 15 min is 8 hours 45 min. Now, subtract the total amount of time, he had off for breaks: 8 hours 45 min − 60 minutes = 7 hours 45 min of actual work time.
f. From 7:30 a.m. until 9 p.m. is 13 hours 30 min. 7 × 13 hours 30 min = 91 hours 210 min = 94 hours 30 min a week.
g. The flight was delayed 15 min, so it will arrive 15 min later than 5:10 p.m., at 5:25.

Review, p. 25

1. a. 6 hours and 52 minutes b. 10 hours and 40 minutes

2. It lands at 6:10 pm.

3. a. 15 min. b. 35 min. c. 38 min. d. 34 min.

4. a. 6 hours b. 14 hours c. 18 min. d. 40 min.

5. It ends at 2:35.

6. The trip was 1 hour 10 minutes long.

Appendix: Common Core Alignment

The table below lists each lesson, and next to it the relevant Common Core Standard.

Lesson	page number	Standards
How Many Hours Pass?	9	3.MD.1
How Many Minutes Pass?	11	3.MD.1
Elapsed Time	14	3.MD.1
More about Elapsed Time	16	3.MD.1
Elapsed Time or How Much Time Passes	20	4.MD.2
Review	25	3.MD.1

Made in the USA
Middletown, DE
02 June 2017